Ana Somarriba
Carlos Jacomino

The New Art of Creating

Ana Somarriba
Carlos Jacomino

The New Art of Creating

How AI Redraws Graphic Design

ScienciaScripts

Imprint
Any brand names and product names mentioned in this book are subject to trademark, brand or patent protection and are trademarks or registered trademarks of their respective holders. The use of brand names, product names, common names, trade names, product descriptions etc. even without a particular marking in this work is in no way to be construed to mean that such names may be regarded as unrestricted in respect of trademark and brand protection legislation and could thus be used by anyone.

Cover image: www.ingimage.com

This book is a translation from the original published under ISBN 978-613-9-46888-1.

Publisher:
Sciencia Scripts
is a trademark of
Dodo Books Indian Ocean Ltd. and OmniScriptum S.R.L publishing group

120 High Road, East Finchley, London, N2 9ED, United Kingdom
Str. Armeneasca 28/1, office 1, Chisinau MD-2012, Republic of Moldova, Europe
Managing Directors: Ieva Konstantinova, Victoria Ursu
info@omniscriptum.com

Printed at: see last page
ISBN: 978-620-8-52789-1

Forewor

Entering *The New Art of Creating: How AI Redraws Graphic Design* is a lot of creativity and technology. This book invites designers, students and creatives to do more than read a book: it is to embark on a journey that redefines the relationship between all disciplines to reflect and act in the face of the digital revolution led by artificial intelligence (AI). What does it mean to be a designer in this new era? Is AI a threat or a tool to expand our creative possibilities?

In its pages, you'll find a rigorous and accessible analysis of how AI is transforming the dynamics of graphic design. From a brief history of its evolution, to concrete tools like Adobe Sensei and DALL-E that are enabling design with greater speed and precision. You will learn how to automate tasks, explore new visual possibilities and customise experiences for specific audiences.

In addition, the book bravely tackles the most challenging questions: How does AI affect human creativity, what role should the designer assume when machines generate options at unprecedented speed, and what is the role of the designer when machines generate options at unprecedented speed? It also reflects on ethical issues, such as copyright and social responsibility, which are essential in an increasingly complex digital environment.

This book is not just a technical guide, it is a manifesto for those who believe that human creativity is still irreplaceable, but can be empowered by collaboration with technology. If you are ready to discover how AI can unlock your creative potential and redraw the future of design, then this book is for you. Welcome to the new age of creation!

Inde

Chapter 1: What is AI and Why does it matter in Graphic Design?

Brief History of AI in Design

Is it possible for a machine to have creativity? This has been a question The use of artificial intelligence (AI) has been recurrent since artificial intelligence (AI) began to intervene in processes previously considered the exclusive domain of humans, such as graphic design.

AI has gone from a futuristic promise to a reality that impacts how we design, automate tasks and democratise access to advanced creative tools. "Artificial intelligence enables systems to exhibit intelligent behaviour by analysing their environment and taking action in the future. measures to achieve specific goals" (Lazo Altamirano, Condori Quispe, & Abarca Rojas, 2024, p. 100). This ability of AI to take on creative functions raises questions and, in turn, opens up unsuspected opportunities in the design industry.

The incursion of AI into design has challenged traditional notions of creativity, redefining the relationship between humans and machines (Abadía, 1986; Lazo Altamirano et al., 2024). This chapter explores how AI is shaping this relationship, while offering a detailed analysis of its practical applications and its ethical and technical impact on graphic design.

Artificial intelligence (AI) is defined as the set of systems and technologies designed to perform tasks that would traditionally require human intervention, including learning, reasoning and problem solving (Pérez & Ramírez, 2023). In graphic design, AI has transcended its role as a technical support tool, positioning itself as a collaborator in creative processes. Its ability to analyse patterns and generate personalised proposals makes it an indispensable resource in a market where speed and innovation are key.

Artificial intelligence (AI) has revolutionised several sectors, and graphic design is at the epicentre of this transformation. Since the 1950s, when the first AI concepts emerged, the technology has evolved from a theoretical blueprint to a practical tool that impacts multiple industries. In graphic design, AI enables both the automation of technical tasks and the generation of visual content at unprecedented speed and scale. AI not only promises to increase efficiency, but also raises profound questions about creativity and the human role in a historically artistic and subjective field (Abadía, 1986; Lazo Altamirano, Condori Quispe, & Abarca Rojas, 2024).

The initial concepts of AI were developed by British mathematician Alan Turing in the 1950s, who suggested that a machine could be considered "intelligent" if it could successfully mimic human responses in communication situations. This idea, known as the Turing Test, laid the foundation for future AI research and enabled systems to be developed that could process and analyse large volumes of data. Today, although AI has not yet reached the level of sophistication and autonomy that Turing envisioned, its applications in visual content generation and design pattern analysis are testament to its value in the graphics industry (Lazo Altamirano et al., 2024).

The journey of AI in graphic design began in the 1960s, when the first computer systems enabled the creation of basic black and white graphics. These early developments represented the foundation for future innovations in digital design. Although the capabilities of these systems were limited, they showed the potential of technology to transform visual creativity (Schwartz, 2021).

Advances in software during the 1990s, such as the launch of Adobe Photoshop and CorelDRAW, marked a fundamental shift in digital graphic design. These tools offered complex edits and advanced image customisation, integrating automatic features such as colour correction and image filters. While these processes were not AI-driven in a modern sense, they laid the foundation for the use of intelligent tools in design (González, 2020).

What Can AI Do in Design?

Today, AI has established itself as a creative partner. Tools such as Adobe Sensei and Runway ML enable the generation of content optimised for specific audiences, adapting to user behaviour and generating graphics, patterns and visual effects that previously required considerable manual work (Adobe, 2022). AI allows designers to experiment without constraints, exploring visual combinations with speed and precision.

This revolution is visible in the use of tools such as DALL-E and Midjourney, which allow designers to generate high-quality images based solely on textual descriptions. A concrete example is the use of DALL-E in advertising campaigns: with a simple instruction, such as "a tranquil city scene at sunrise, with a dreamy tone", the system can create multiple options that meet this description. This means that the designer can quickly visualise different concepts without the need to create manual sketches or spend hours in editing programs. According to Santos Tapia (2024), AI facilitates "rapid visual experimentation without the high time costs that characterise manual design" (p. 87). This speed and adaptability have made AI an indispensable tool for modern designers seeking to improve efficiency without sacrificing quality.

Today, AI not only supports automation, but also drives visual innovation. From generating complex graphics in seconds to customising designs for different audiences, this technology has revolutionised designers' workflow, allowing them to explore new creative frontiers while optimising time and resources (Wilson, 2023).

One of the most common and practical uses of AI in graphic design is in image editing. AI algorithms can now identify and correct specific elements in an image, such as adjusting colour, lighting and contrast automatically. Tools such as FaceApp and Photoshop have implemented algorithms that, using convolutional neural networks, can retouch faces, remove backgrounds and restore images with astonishing accuracy (González, 2020). This capability not only saves time, but also allows advanced edits to be made without requiring extensive technical skills.

However, the AI revolution in graphic design is not limited to image generation. Another innovative aspect is its ability to optimise and customise visual elements in real time. Adobe Sensei, the AI integrated in Adobe Creative Cloud, allows for automatic adjustments of lighting, colour and contrast in an image, and even suggests font and style combinations, helping designers make informed aesthetic decisions in seconds (Martínez, Dennis, Cáceres, Gutiérrez, & Quiroz, 2023). This level of automation is crucial in fast-paced industries, such as digital advertising and social media, where turnaround times are short and quality must be high.

Beyond automation, AI acts as a 'creative assistant', suggesting ideas and solutions based on historical data and current trends. This helps designers overcome creative blocks, exploring new visual combinations quickly and efficiently (Taylor & Green, 2022).

In addition, AI offers the possibility to tailor designs to specific audiences. In the past, designers had to create manual variations for different audiences, which was a time-consuming and laborious process. Today, thanks to AI, the same piece can be automatically personalised for different audience segments, a process known as 'adaptive graphic design' (Lazo Altamirano et al., 2024). This approach allows each design to adapt to the colour, style and composition preferences of its target audience, maximising the impact and reach of visual communication.

It also allows designers to generate custom graphics, logos and patterns automatically. Tools such as Canva and Looka use algorithms to analyse visual preferences and create logos and graphics tailored to the specific styles of each user (Rosenberg, 2021). These programmes automate the design process, generating multiple visual options that are aligned with traditional design principles, but tailored to the user's preferences, saving time and stimulating creativity.

Artificial intelligence is defined as a discipline of computer science that focuses on creating systems capable of performing tasks that, in the past, could only be performed by humans.
Although today's AI lacks the "consciousness" or "creativity" that characterises the human intellect, its ability to analyse data, learn from patterns and generate complex solutions has profoundly transformed many fields, including graphic design. In the words of Abadía (1986), AI has
"consists of systems that exhibit characteristics associated with human intelligence, such as learning, reasoning and problem solving" (p. 1). The evolution of this technology has allowed it to become an everyday tool for millions of people, extending its applications to industrial, educational and creative levels.

Graphic design has benefited greatly from advances in AI, not least because of its ability to democratise advanced tools that were previously only available to experts. For example, platforms such as Canva and Crello offer inexperienced users the ability to design engaging graphics using AI-assisted features that optimise colour, font and style combinations. Instead of requiring advanced knowledge, users can select predefined options that have been optimised by AI to meet aesthetic and functional criteria. According to Santos Tapia (2024), this accessibility "transforms graphic design, making it accessible and adaptable for anyone, regardless of their background" (p. 85).

In addition to democratising design, AI has raised new questions about the value of graphic design in a context where anyone can create high-quality visual content. Where does

the role of the professional designer fit in a world where AI tools make graphic design easier for the everyday user? While some in the industry fear that AI could displace designers, others believe that AI complements and enhances their skills rather than replacing them.
Martinez et al. (2023) note that "AI allows designers to focus on creativity and message, leaving repetitive tasks to algorithms" (p. 518), which redefines the role of the designer in a digital age.

One of the most fascinating aspects of AI in graphic design is its ability to transform every stage of the creative process, from initial inspiration to final production. In the past, a designer might spend days or weeks experimenting with different colour combinations, typefaces and compositions. Today, AI allows much of this work to be done automatically, significantly speeding up the workflow and facilitating visual experimentation (Santos Tapia, 2024).

Far from displacing the designer, AI is emerging as a creative collaborator. According to Wilson (2023), AI should be understood as a 'catalyst for creativity', providing suggestions, automating repetitive tasks and expanding visual possibilities. Tools such as Adobe Sensei do not replace human intervention, but rather enhance the designer's ability to experiment with ideas that previously would have been unfeasible due to time or resource constraints.

A less explored but equally fascinating aspect is the role of AI as a source of inspiration for designers. Rather than simply generating final content, AI can suggest initial ideas that serve as a starting point for more elaborate projects. According to Taylor and Green (2022), AI tools act as 'creative assistants', helping designers overcome creative blocks by generating innovative proposals based on current trends and historical data.

A recurring question is whether AI can be considered creative in a human sense. According to Pérez and Ramírez (2023), algorithmic creativity is based on the combination of existing patterns and data, which makes it possible to generate original but not necessarily innovative results in human terms. For example, tools such as DALL-E can produce unique images based on textual descriptions, but its creativity is limited by the data it was trained on.

True creativity, argue Veritas (2023), lies in the designer's ability to interpret and refine AI-generated proposals, adding a cultural and emotional context that machines cannot understand.

AI has made graphic design more accessible than ever before. Platforms such as Canva allow non-technical people to create professional graphics using AI-assisted features. These tools democratise design, but also raise questions about the role of the professional designer.

While this accessibility may lead to increased competition, it also offers an opportunity for designers to differentiate themselves through quality and innovation. According to Fernandez and Lopez (2023), designers who embrace AI as an extension of their skills can offer unique value in a market saturated with generic content.

Despite its advances, AI has significant limitations. It lacks intuition, empathy and cultural context, essential skills for creating designs that resonate with audiences. According to Rodriguez (2022), AI can generate impressive visual solutions, but these must be critically evaluated by the designer to ensure they meet the project's objectives.

In addition, inherent biases in AI training data can negatively influence outcomes, generating content that is not inclusive or perpetuates stereotypes (Brown, 2023). This underscores the importance of designers acting as critical mediators between AI propositions and client needs.

As AI becomes more deeply integrated into graphic design, the role of the designer is evolving. According to Wilson (2023), designers of the future will need to combine traditional artistic skills with advanced technical competencies. This includes understanding how algorithms work, interpreting the results generated by AI, and collaborating with technology experts to develop innovative solutions.

The future of AI in graphic design promises exciting developments, from integration with emerging technologies such as virtual reality to the development of more intuitive systems that actively collaborate with designers. According to Lopez and Garcia (2024), these innovations will not only expand creative possibilities, but also redefine how we think about the design process. For example, AI is expected to evolve into systems capable of better interpreting emotions and cultural contexts, which could enable the creation of more empathetic and communicatively effective designs.

Objectives and Purpose of the Book

Therefore, this book seeks to answer an essential question for the digital age: How is artificial intelligence revolutionising graphic design and what will its impact be in the future? Through a rigorous analysis of current AI tools and applications, readers will gain an in-depth understanding of how this technology influences design and the creative process. This study will also address the ethical and technical challenges that accompany AI in the industry, offering a critical perspective on how designers can adapt to an increasingly automated world without losing their creative touch.

Artificial intelligence has redefined graphic design by combining technical efficiency with creative possibilities never seen before. Far from being a threat, AI is positioned as a strategic ally that allows designers to expand their capabilities and approach projects with a more innovative approach. However, to take full advantage of this technology, it is essential that design professionals adopt a balanced approach, integrating unique human competencies such as empathy and storytelling with the analytical and automation capabilities of AI.

According to Veritas (2023), "AI does not replace human creativity, but rather amplifies it, providing tools that empower the imagination and allow for the exploration of visual territories that were previously unattainable" (p. 87). Therefore, the future of graphic design lies in symbiotic collaboration between humans and machines, where each brings its strengths to build a more dynamic, inclusive and sustainable field.

Chapter 2: AI in Action: Tools and Practical Examples

Artificial intelligence (AI) has transformed graphic design by offering innovative tools that not only optimise processes, but also inspire new creative possibilities.

In recent years, artificial intelligence (AI) has revolutionised graphic design by introducing tools that optimise, personalise and improve creative processes. AI in design not only facilitates the creation of visual content, but also enables large-scale personalisation and the adaptation of advertising campaigns in real time. This chapter explores how AI is applied in graphic design through key tools such as Adobe Sensei, DALL-E and Midjourney, as well as in virtual reality projects and advertising campaigns, transforming the creative industry and offering previously unimaginable possibilities (Sánchez, 2024; Chaparro González, 2024; Jiménez Sánchez, 2024).

Main AI Tools in Design

Adobe Sensei: Adobe Sensei is an AI platform from Adobe that powers applications such as Photoshop and Illustrator, allowing designers to automate complex tasks and optimise workflows. Sensei offers features such as Content-Aware Fill, which can remove elements from an image and fill the empty space with contextual content in a matter of seconds, saving time in manual editing (Adobe, 2022). This technology analyses visual patterns and proposes changes that improve composition, helping designers to focus on the creative aspect (Jones & Smith, 2022, p. 78).

Runway ML: Multimedia Content Generation

Runway ML is an AI tool designed for graphic designers and videographers. This platform allows the generation of animations, videos and visual effects from simple inputs, such as text or images. According to Santos Tapia (2024), "Runway ML offers designers the ability to transform complex ideas into professional visual content in minutes" (p. 92). This is especially useful when creating content for advertising campaigns and digital platforms.

Wix ADI: Wix ADI (Artificial Design Intelligence) is a website creation tool that allows users to design custom pages without prior programming experience. By answering a few initial questions, Wix ADI uses AI to organise content, select colour palettes and adjust the design to suit the user's needs (Lazo Altamirano et al., 2024). This results in unique and functional sites that respond to both user style and visitor expectations (Rosenberg, 2021, p. 23).

Canva: Canva has integrated AI into its platform to facilitate fast and professional design. Features such as "Magic Resize" allow designs to automatically adapt to different sizes for social platforms, and AI suggests colours and fonts that match the user's visual style (Martinez & Torres, 2022). These capabilities allow non-designers to create consistent and attractive graphics for social networks and other media (Martinez & Torres, 2022, p. 87).

Practical Usage Example: A designer working on an advertising campaign can enter a description such as "an urban sunset scene with a futuristic feel and warm colours". In a matter of seconds, the AI will generate several versions of the image that meet these criteria, allowing the designer to explore different concepts without spending hours on sketches or manual revisions. According to Sanchez (2024), this speed and adaptability make AI an indispensable tool in the industry.

One of the most notable innovations is AI-driven typography creation, where visually harmonious font combinations are automatically generated. Tools such as Fontjoy use AI to analyse the aesthetics of each font and propose combinations that reflect the identity of a brand or project (Nguyen et al., 2019). This technology helps designers save time in typographic selection and achieve visual consistency efficiently (Schwartz, 2021, p. 40).

While digitisation dominates graphic design today, AI is also significantly impacting print and editorial design. Tools such as InDesign AI integrate algorithms that analyse content to propose balanced and aesthetically pleasing layouts, improving readability and visual layout (Rosenberg, 2021). These applications automate tasks such as font selection, text alignment and the creation of consistent styles throughout a project.

AI is helping designers anticipate social media trends, generating content that resonates with audiences before topics go viral. Tools such as BuzzSumo AI analyse interaction data and suggest visual content types that might attract more attention, such as predominant colours or platform-specific formats (Santos Tapia, 2024).

The proliferation of mobile devices has prompted the development of AI tools that automatically optimise layouts for different screen sizes. This includes adjustments to the layout of text, images and interactive buttons, improving the user experience . Tools such as Sketch AI can generate mobile versions of a web design in a matter of minutes (Rosenberg, 2021).

AI has taken personalisation to a level never seen before, allowing companies to tailor their advertising campaigns to specific audiences. By analysing consumer behavioural data and preferences, AI can personalise visuals and messages in real time, adjusting elements of colour, style and composition according to audience preferences.

This, known as "adaptive design", allows the same campaign to be adapted to multiple audience segments, maximising the impact and relevance of each visual piece (Lazo Altamirano, Condori Quispe, & Abarca Rojas, 2024).

Jiménez Sánchez (2024) analyses how personalisation in advertising campaigns not only improves the effectiveness of the message, but also allows brands to connect emotionally with consumers. AI allows visual content to automatically adjust to the interests of the audience, thus improving campaign reach and user engagement. For example, a fashion brand could tailor its ads according to the style preferences of its consumers, using AI to personalise each visual based on the prevailing trends in each market.

Example of a Successful Campaign: Nike has implemented AI in its ads to adjust visual elements according to the target market. This strategy has allowed its campaigns to vary in colour, typography and style, depending on the target audience, optimising the emotional connection between the brand and consumers and increasing the conversion rate (Jiménez Sánchez, 2024)

Graphic design in virtual reality (VR) has also benefited from AI tools, which optimise development processes and reduce costs without compromising visual quality. Chaparro González (2024) explains that AI can significantly speed up the creation of virtual reality environments by automating tasks that previously required hours of manual work. By applying AI to VR projects, designers can explore visual concepts quickly and efficiently, generating demos and interactive experiences that take graphic design to a more immersive level.

Example Application: A design studio working on a virtual reality demo can use AI to create a virtual environment based on a description, such as "an enchanted forest at dusk with a mystical atmosphere". The AI automatically generates visual elements, textures and lighting effects that are integrated into the VR experience. According to Chaparro González (2024), "AI tools allow optimising the development process, reducing costs and time without compromising graphic quality" (p. 22).

Examples of AI in Different Types of Design

One of AI's most valuable capabilities is its ability to analyse large volumes of data and predict behavioural patterns. According to Martínez and Torres (2022), AI algorithms allow designers to anticipate user preferences, resulting in more effective campaigns and visual content tailored to specific audiences.

AI has also expanded the scope of graphic design, making it more accessible and inclusive. AI tools can automatically generate descriptions of images, allowing people with visual impairments to understand visual content. In addition, AI can adjust design elements to make them easier to interpret for people with cognitive disabilities, creating a more inclusive and accessible design environment for all (Lazo Altamirano et al., 2024).

While graphic design has evolved digitally, AI is also transforming traditional print and editorial design processes. Tools such as PageMaker AI use algorithms to optimise page layout in magazines, books and catalogues. These tools analyse content to propose visually balanced layouts, optimising space and improving readability (Rosenberg, 2021).

Visual storytelling is being revolutionised by AI by enabling the creation of content that tells stories dynamically. Tools such as Storyboard AI generate visual sequences from textual scripts, suggesting cinematic compositions and approaches. According to Jiménez Sánchez (2024), "this technology accelerates the processes of narrative content creation, allowing designers to focus on more artistic aspects" (p. 84).

Generative design is one of the most promising trends at the intersection of AI and graphic design. This approach allows designers to set basic parameters, such as colours, shapes and styles, while AI generates multiple variations within those boundaries. According to Chen (2022), generative design not only accelerates creative processes, but also opens up new possibilities to explore innovative solutions that might not have been considered by traditional methods.

A prominent example is the use of AI to design corporate logos. Tools such as Looka generate hundreds of options from a simple description, allowing companies to select and refine the design that best suits their visual identity (Gómez & Rodríguez, 2023).

Advertising and Personalisation: AI in Digital Marketing

Personalisation is key in digital marketing, and AI has enabled an unprecedented level of precision in advertising campaigns. By analysing user behavioural data, AI adjusts ads to show relevant content at the ideal time, maximising advertising impact (Sanchez, 2024).

AI allows brands to create ads tailored to each user's preferences, improving the experience and increasing the likelihood of conversion. This type of personalisation is particularly effective in remarketing campaigns, where products that the user has shown interest in in the past are presented (Santos Tapia, 2024). By adjusting ads in real time, AI is able to maintain relevance and improve click-through rates (Gomez & Rodriguez, 2023, p. 18).

According to Santos Tapia (2024), "AI can transform graphic design into a more inclusive medium, allowing designers to create content that is accessible and understandable to a more diverse audience" (p. 92). This inclusive aspect of AI in graphic design reinforces the purpose of design to communicate effectively and meaningfully, without excluding any group of people.

Graphic design does not work in isolation, and AI is facilitating collaboration across disciplines. For example, in the development of mobile apps or video games, graphic designers work alongside developers and user experience (UX) experts. Tools such as Figma, which now integrate AI-based functions, enable real-time interactive prototyping. According to Martínez and Gómez (2022), this efficient collaboration significantly improves delivery times and the quality of the final product.

Artificial intelligence has proven to be a game changer in graphic design, facilitating processes, opening up new creative possibilities and improving accessibility. From tools such as Adobe Sensei to innovative applications in virtual reality and generative design, AI is transforming the way designers work and collaborate. According to López and García (2024), "true innovation in design lies not in replacing the designer, but in empowering their creativity and pushing the boundaries of what is possible" (p. 45).

In the field of animation, AI is revolutionising creative processes. Tools such as DeepMotion make it possible to generate animations from videos, using deep learning to interpret human movement. This not only saves time, but also allows graphic designers to explore new ways of visual storytelling (Wilson, 2023).

The concept of dynamic visual identity has gained prominence thanks to AI. Tools such as Dynamic Brand AI allow brands to automatically adapt their logos, colours and graphic elements to different platforms and contexts. For example, a logo can be modified to include specific colours to match the current event or season, without the need for a manual redesign.

Gómez and Rodríguez (2023) explain that "this ability to adapt graphics in real time improves the brand's perceived relevance and strengthens its connection with the public" (p. 43).

IA in Web Design and User Experience

AI in web design improves navigation through recommendation systems and internal search engines that display relevant content for the user. In addition, websites adapt their design to devices and environments, ensuring optimal viewing on mobile phones, tablets and computers (Chaparro González, 2024). This adaptive design ensures that the user has a smooth and consistent experience on any platform. Finally, AI has taken adaptive design to a higher level, automatically adjusting visual elements and functions according to the user's behaviour and context. This ability to adapt in real time improves both the aesthetics and functionality of websites, allowing for a more personalised and optimised experience (Schwartz, 2021).

Artificial intelligence has redefined graphic design, integrating deeply into creative, technical and strategic processes. From automating tasks to creating innovative visual content, AI not only improves efficiency, but also expands creative possibilities. According to López and García (2024), "AI does not merely simplify processes, but inspires new ways of conceiving and executing projects, opening an era of collaborative creativity between humans and machines" (p. 45). The key to success lies in how designers adopt and adapt these tools, using them to enhance their artistic vision and meet the demands of an increasingly connected and demanding world.

Chapter 3: What Changes for the Designer?

Artificial intelligence (AI) is transforming the way we work, especially in the field of design. This chapter focuses on how AI impacts the work of designers, the skills that become crucial in this new era, and examples of collaboration between humans and AI. As we explore these topics, it will reveal how AI not only automates tasks, but also empowers creativity and innovation.

The integration of AI in graphic design has repositioned the designer as a mediator between technological capabilities and human needs. According to López and García (2024), "According to López and García (2024), the cultural impact of AI in graphic design positions the designer as a mediator between technology and human audiences, especially in emotionally relevant contexts" (p. XX). This implies that designers must not only master technological tools, but also deeply understand their users in order to generate meaningful experiences.

With the advent of artificial intelligence (AI), the role of the graphic designer has evolved beyond traditional technical and creative tasks to a more strategic and consultative approach. According to Wilson and Green (2023), AI does not replace the designer, but rather redefines the designer's job by automating repetitive tasks and freeing up time for conceptualisation and visual strategy development.

With the growth of AI tools, designers now take on broader roles in digital project management. According to Hernández and Pérez (2023), "AI facilitates workflow automation, but it also requires strategic planning to integrate these tools into multidisciplinary creative teams" (p. 58).

Task Automation: What does AI do for us?

For example, a designer who previously spent hours adjusting colour and lighting in a composition can delegate these tasks to tools such as Adobe Sensei, while concentrating on the visual narrative and emotional connection of the design (Academy Partners, 2023). This transforms the designer into a creative facilitator who works alongside machines to optimise processes and generate innovative solutions.

Technological change demands a constant updating of skills for graphic designers. According to López and Hernández (2024), professionals must incorporate technical competencies, such as understanding algorithms and AI models, along with human skills, such as empathy and contextual creativity.

Beyond their technical skills, the designer of the future will be a strategist capable of leading complex projects that integrate creativity, technology and business objectives. According to McKinsey (2023), companies that manage to incorporate designers in strategic roles alongside AI will be able to generate a significant competitive advantage in their markets.

The automation of routine tasks is one of the most significant contributions of AI to design. AI tools can handle repetitive activities such as image editing, creating initial sketches and organising files, allowing designers to focus on more creative and strategic aspects. According to a study by Deloitte, organisations that implement AI can save up to two hours a day on mundane tasks, resulting in a 35% increase in time spent on creative activities (Deloitte, 2020).

AI can analyse large volumes of data to identify trends and patterns that might go unnoticed by a human. For example, tools such as Adobe Sensei use algorithms to suggest colour combinations and styles based on user preferences and current trends (Academy

Partners, 2023). This not only improves the efficiency of the creative process, but also offers valuable insights that can inform design decisions.

Furthermore, automation is not only about technical tasks; it also includes content generation. AI can create variations of an existing design or propose new ideas based on parameters defined by the designer. This allows multiple options and approaches to be explored quickly without the wear and tear associated with manual work (Cely & Buake, 2023).

With less time spent on repetitive tasks, designers have the opportunity to focus on strategic creativity. This involves thinking about how their designs can influence user behaviour and how they align with broader business goals. The ability to collaborate with AI tools allows designers to experiment with new ideas and concepts without fear of wasting valuable time (AI Tools, 2023).

While AI has introduced unprecedented levels of automation, it also raises questions about the creative autonomy of the designer. According to Veritas (2023), AI amplifies human creativity by providing tools that enhance imagination (*Creativity in the AI age: Challenges and opportunities*, p. XX).

Strategic creativity, combining commercial goals and aesthetic values, has become a central focus for designers in the age of AI. According to Miller et al. (2023), AI tools provide data-driven insights that inform design decisions, ensuring that creations are not only appealing, but also effective in meeting specific goals.

Skills Gaining Relevance in the Age of AI

As designers begin to work alongside AI systems, certain skills become essential to maximise this collaboration. These include:

- Critical Thinking: The ability to critically evaluate the results generated by AI systems is critical. Designers must be able to discern when an AI-generated suggestion is appropriate or when it needs adjustment (Cely & Buake, 2023)
- Adaptive Creativity: Although AI can generate ideas based on existing data, human creativity remains unique. Designers must learn to use AI as a tool to enhance their creativity, adapting their approaches according to what the technology offers them (Veritas, 2023)
- Interdisciplinary Collaboration: Working effectively with multidisciplinary teams is crucial. Designers must communicate clearly with engineers and data experts to ensure successful projects (Academy Partners, 2023)
- Continuous Learning: The rapid evolution of technological tools requires designers to be willing to constantly learn and adapt. This includes becoming familiar with new AI-driven design platforms and understanding their capabilities and limitations (AI Tools, 2023)

These skills are not only needed to interact with AI tools but also to lead creative projects where technology and design are intertwined.

Collaboration between humans and AI has led to surprising results in a variety of design projects. One notable example is the use of generative AI in architectural design. Firms such as Zaha Hadid Architects have used algorithms to explore new architectural forms that would be difficult to conceive manually (LHH, 2023). This combination makes it possible to create innovative and functional structures by merging human judgement with the advanced analytical capabilities of AI.

Another case is the use of tools such as Runway ML in graphic design. This platform allows designers to generate images or videos from textual descriptions, thus speeding up the creative process and opening up new aesthetic possibilities (Cely & Buake, 2023). This approach not only improves the final product but also fosters a collaborative environment where both humans and machines learn from each other.

In digital marketing, companies are using AI algorithms to personalise advertising campaigns based on consumer behaviours. This allows designers to create more relevant and engaging content for their audience (Academy Partners, 2023). The synergy between humans and AI not only optimises processes but also enhances human creativity.

Examples of Collaboration

The interaction between designers and artificial intelligence (AI) tools is not only redefining creative processes, but also pushing the boundaries of what is possible in terms of innovation. As technology continues to evolve, emerging areas where AI plays a crucial role have been identified, creating new opportunities and challenges for design professionals.

One of the most promising applications of AI is its ability to enable personalised design on a large scale. For example, platforms such as Canva have integrated AI-powered tools that allow users to customise templates automatically based on their specific preferences and needs (Lee et al., 2022). This significantly reduces the time needed to create content and allows designers to focus on more strategic projects.

Virtual and augmented reality technologies, powered by AI, are transforming experience design. For example, Unity and Unreal Engine integrate AI algorithms that enable the creation of detailed and dynamic virtual environments in record time (Miller et al., 2023). These tools are particularly useful in creating immersive user experiences that seamlessly combine the physical and the digital.

In the field of industrial design, AI is accelerating prototyping processes. Tools such as Autodesk Fusion 360 allow designers to create iterative models quickly using AI-based simulations (Johnson, 2023). This not only saves time and resources, but also encourages the exploration of multiple solutions before settling on a final design.

The increasing integration of AI into design processes is changing the structure of creative teams. According to a McKinsey report (2023), designers are expected to take on more strategic and consultative roles, delegating technical tasks to advanced AI systems. This shift emphasises the importance of human skills such as empathy, storytelling and context-based decision-making.

Human-machine collaboration is fostering a new model of collective creativity. Projects such as Google's DeepDream have demonstrated how AI tools can inspire designers to explore aesthetics and concepts that would otherwise be difficult to imagine (Kaur, 2023). This collaborative approach is redefining the role of the designer as a facilitator of creativity.

Artificial intelligence does not replace designers, but amplifies their ability to solve complex problems and create innovative solutions. However, its successful integration depends on a careful balance between human skills and technological capabilities . As Veritas (2023) noted, "the designers who will thrive in this era will be those who understand how to use AI as an extension of their own creativity".

The role of the designer is evolving towards a more strategic and consultative approach. According to McKinsey (2023), AI is destined to take over technical tasks, while designers focus on areas such as visual storytelling, empathy and contextualised decision-making. This shift highlights the importance of combining unique human skills with advanced technological capabilities.

The advent of artificial intelligence in graphic design does not mark the end of human creativity, but the beginning of a symbiotic collaboration between designer and machine. In this new paradigm, AI acts as a tool that enhances the human ability to solve complex problems and explore previously unattainable creative possibilities. However, as Taylor and Green (2022) point out, "the real value of graphic design lies not in automation, but in the designer's ability to give meaning and context to the results generated by AI" (p. 84).

This balance between technology and creativity requires designers to evolve, adopting more strategic roles and developing critical skills that ensure that technological tools amplify, not dilute, their artistic vision. In the age of artificial intelligence, the designer is not only a creator, but also a curator of visual experiences that connect deeply with audiences in an ever-changing world.

The challenge for designers lies in learning to balance automation with intuition, using AI to enhance, not replace, their work. In addition, it is crucial that they take ethical responsibility in the use of these technologies, ensuring that the results are inclusive, original and culturally relevant.

Chapter 4: Ethical Challenges and Issues

The integration of artificial intelligence (AI) into graphic design has not only revolutionised the we create images, typography and layouts, it has also opened up a world of possibilities and dilemmas that few could have anticipated. As machines begin to play a central role in human creativity, fundamental questions arise: Who is the true author of an AI-generated design? How will this affect the future of designers? And most importantly, how can we ensure that the technology is used ethically and responsibly?

This chapter explores these issues from an innovative and thoughtful perspective, highlighting the legal challenges, labour market transformations and ethical considerations that designers must take into account. In a world where technology is evolving faster than the regulations that govern it, it is crucial to understand the challenges and opportunities that AI offers, not just as a tool, but as a transformative agent for the graphic design industry.

The integration of artificial intelligence (AI) into graphic design has not only revolutionised the way we create images, typography and layouts, but has also generated ethical dilemmas and profound transformations in the industry. In a world where technology is advancing faster than the regulations that govern it, it is crucial to analyse how designers can adapt to lead ethically and innovatively.

The advent of AI has led many to question what it means to be creative. According to López and García (2024), AI does not generate ideas from scratch, but rather combines existing patterns and data. This implies that, although the results may seem original, they are essentially the product of a collection and recombination of previous information. This phenomenon raises a key question: is creativity exclusive to humans or can machines engage in meaningful creative processes?

In this sense, the role of the designer as interpreter and curator becomes more important than ever. Designers must not only evaluate the technical quality of AI-generated outputs, but also contextualise and adapt them to align with cultural, emotional and commercial values. According to Hernandez and Torres (2024), "AI extends the designer's tools, but it does not replace the human capacity to give meaning and purpose to design" (p. 45).

A fundamental aspect to be addressed in this context is how AI redefines the concept of creativity in graphic design. According to López and García (2024), AI does not create from nothing, but builds on existing data and patterns. This raises the question of whether AI-generated design can be considered truly creative or simply a reflection of the knowledge accumulated in its databases. This technical limitation underlines the importance of human designers as interpreters and contextualisers of AI-generated proposals.

Impact of AI on the Design Labour Market

The design job market is undergoing an unprecedented transformation due to the incorporation of AI tools. These technologies have the potential to automate repetitive tasks, such as colour selection, basic image editing or template creation, freeing designers to focus on more strategic and conceptual aspects of their projects (Fernández & López, 2023). However, this evolution also brings with it new challenges.

The impact of AI on graphic design is not only technical or economic, but also emotional. Many designers are anxious about the possibility of being replaced by automated tools, especially in tasks that previously required specialised skills (Johnson, 2023). According to a study by Lee and Park (2023), 62% of designers interviewed expressed concern about how AI could reduce the demand for their work. This perception may generate resistance to the adoption of new technologies, slowing down their integration into the industry.

On the other hand, AI can also generate emotional benefits by freeing designers from repetitive tasks, allowing them to focus on creativity and conceptualisation. According to Taylor and Green (2022), this 'symbiotic collaboration' can increase job satisfaction by giving designers more time to innovate and experiment.

Moreover, AI-generated creativity is based on pre-existing patterns, trained from data collected from human works. This means that, although the result may appear original, it is actually a combination of existing references (Chen, 2022). This phenomenon raises important questions about originality and the role of designers in a future dominated by machines capable of "creating".

The entry of AI into design not only redefines the skills required, but also forces a rethink of the role of the designer in an increasingly dynamic and competitive labour market. Key aspects are highlighted below:

1. **Polarisation of Skills and Roles**: While designers who master emerging technologies have a competitive advantage, those who are limited to traditional skills face risks of job exclusion. Investment in lifelong learning should not only be individual, but also institutional, promoted by governments and companies to reduce the technology gap (Johnson, 2023).

2. **Intensified Global Competition**: Accessible AI-based tools have enabled people without formal design training to compete in freelance markets, lowering costs but also affecting the perceived value of professional design (Martinez & Gomez, 2022).

3. **New Job Opportunities**: AI not only automates tasks, but also creates new areas of work, such as managing creative algorithms or curating AI-generated content, which require hybrid skills between creativity and technology (IDA, 2023).

Ethics and Responsibility in Graphic Design

Ethics in AI-based graphic design encompasses critical issues such as authorship, transparency and cultural preservation:

1. **Authenticity and Transparency**: The use of AI should be clearly communicated to clients and the public. Reporting whether a work was created or assisted by AI builds trust and avoids legal or perception conflicts (Taylor & Green, 2022).

2. **Preserving Cultural Diversity**: While AI can amplify cultural styles, there is also a risk of homogenising traditions. Responsible designers must actively work to ensure that technological tools reflect cultural richness, rather than dilute or stereotype it (Chen, 2022).

3. **Prevention of Manipulation and Disinformation**: AI-generated images can be used in disinformation campaigns. Therefore, it is essential to implement clear ethical standards and controls to prevent the misuse of these technologies (Rodriguez, 2022).

One of the main effects of AI on the labour market is the polarisation of roles. While designers with advanced skills in technology and programming have greater opportunities, those with a more traditional approach may face difficulties in staying competitive (Johnson, 2023). This phenomenon creates a gap in the industry that could widen if there is no investment in lifelong learning.

Moreover, the democratisation of AI-based tools has allowed non-professional users to access advanced design technologies. This has generated intense competition, especially in the freelance field, where clients are looking for quick and cost-effective solutions (Martínez & Gómez, 2022). While this accessibility encourages creativity, it can also saturate the market with lower quality work.

However, it's not all negative. AI is also redefining what it means to be a graphic designer. According to the *International Design Association* (IDA, 2023), designers of the future will need to adopt a hybrid approach that combines traditional artistic skills with advanced technological competencies. This includes the ability to work with algorithms, understand machine learning and collaborate with engineers to develop innovative solutions.

Finally, AI can help improve inclusion in design. Tools such as Adobe Sensei, which uses AI to optimise designs, can facilitate the work of people with disabilities, allowing them to participate in a field that was previously inaccessible to them (Wilson, 2023). This development highlights the potential of AI to not only transform the industry, but also make it more inclusive and diverse.

Copyright and Intellectual Property Rights in AI Design

The adoption of AI in design has also brought with it ethical considerations. According to López and García (2024), designers must be aware of how their decisions affect privacy, equity and cultural representation in their projects.

The ethical impact of the use of AI in graphic design is an issue that cannot be ignored. AI tools have immense power to influence public perception, but they can also be misused to manipulate or misinform. For example, the creation of hyper-realistic images through AI has raised concerns about their use in disinformation and propaganda campaigns (Rodriguez, 2022). This problem underlines the importance of establishing clear ethical standards for the use of these technologies.

The lack of clear regulations for AI in graphic design poses a significant challenge. According to Smith and O'Connor (2023), the absence of international standards makes it difficult to protect the rights of designers, especially in global markets. This has led to debates about the need for an international ethical charter, which would regulate the use of AI in creative industries. Such a charter could establish minimum criteria for transparency, attribution of authorship and responsible use of these tools.

Analyse how AI uses pre-existing images and designs to train generative models, which can lead to legal disputes related to copyright. According to Pérez and Ramírez (2023), current laws do not adequately address the question of whether designers whose work is used to train AI should receive compensation or recognition.

Case study: In 2023, several artists sued AI platforms for training their models with copyrighted works, arguing that the tools replicated their styles without proper attribution

(Chen, 2022). This example underscores the need for clear regulations that balance the rights of creators and technology companies.

Transparency is another crucial aspect. Designers have a responsibility to inform their clients and the public when using AI in their projects. According to Taylor and Green (2022), this practice not only builds trust, but also helps educate consumers about the capabilities and limitations of the technology.

Training is key for designers to adapt to a transformed labour market. According to Lee and Park (2023), 78% of designers interviewed in a global study agreed that AI-related skills will be essential in the next five years. However, unequal access to these training opportunities could exacerbate economic and social gaps in the industry.

Discuss how AI tools can be used for unethical purposes, such as generating misleading content or manipulating audiences through misinformation. According to Rodriguez (2022), this poses significant challenges in the graphic design industry, especially in the creation of hyper-realistic images that could be used in propaganda campaigns.

In addition, it is essential to consider the cultural impact of AI on design. While these tools can generate works inspired by diverse styles, they also run the risk of homogenising cultural traditions. According to Chen (2022), it is vital that designers actively work to preserve cultural diversity in their projects, using AI as a means to amplify, not replace, cultural expressions.

Universities and educational platforms are beginning to include AI in their curricula, offering courses that combine artistic fundamentals with technological knowledge . This not only benefits designers, but also prepares the next generation of professionals for a hybrid labour market (Smith & O'Connor, 2023).

One of the biggest debates surrounding AI in graphic design is the question of authorship. If an AI tool generates a design, who is the legal author? According to Pérez and Ramírez (2023), current intellectual property laws do not adequately address this scenario, as they were designed to protect human creations. This has led to conflicts over the rights to use and distribute AI-generated works, especially in commercial contexts.

A landmark case occurred in 2022 when an artists' collective sued a company for using AI models that replicated their styles without compensation or proper attribution. This case highlights the need to update regulation to protect both designers and companies using AI (Chen, 2022).

Finally, designers should reflect on the purpose behind the use of AI. Questions such as "Is this tool improving the project?" or "Is it ethical to automate this task?" are essential to ensure a responsible approach. According to the *Ethical Design Initiative* (EDI, 2023), ethics in design should not be a secondary consideration, but a fundamental pillar of professional practice.

Explore how working with AI affects designers' perceptions of their own creative value. According to Hernandez and Torres (2024), while AI can free designers from repetitive tasks, it can also lead to insecurity about whether their work is valued in an increasingly automated environment.

Proposition: Promoting the idea that AI does not replace the designer, but rather amplifies their ability to generate impact, can alleviate some of this anxiety and foster a mindset of collaboration rather than competition (Taylor & Green, 2022).

The integration of AI into graphic design is not only transforming the way images, typography and layouts are created, it is also reshaping the ethics, economics and culture of the industry. Designers have the opportunity to lead this change by adopting responsible practices, investing in continuing education and actively collaborating with technology to address emerging challenges. By prioritising transparency, sustainability and cultural diversity, AI can become a tool for creating more inclusive, ethical and innovative graphic design.

The integration of artificial intelligence in graphic design represents a historic turning point in the way we conceive creativity and visual production. While AI has proven to be a powerful tool for optimising processes and expanding creative possibilities, its implementation also requires deep reflection on ethical responsibilities, sustainability and cultural impact. According to López and García (2024), "the true value of AI lies not only in what it can do, but in how we choose to use it responsibly and meaningfully" (p. 58).

As technology advances, it is crucial that designers do not lose sight of their role as cultural and ethical mediators. AI should be seen as a collaborator, not a substitute, that enhances human skills and enriches visual storytelling. The future of graphic design will depend on how professionals manage to integrate these tools to overcome barriers, foster diversity and preserve the human sense in every creative project. In the end, AI does not define creativity; it is human decisions that give purpose and meaning to what we do.

Chapter 5: The Future of Graphic Design in the Age of AI

The emergence of artificial intelligence (AI) in graphic design is redefining not only creative processes, but also the role of designers in a constantly evolving visual ecosystem. Beyond automation, AI has become an essential tool for exploring new forms of creativity, adapting to the needs of the market, and adapting to the needs of the market.
market demands and respond to today's social and ethical challenges.

The integration of artificial intelligence (AI) into graphic design is profoundly reshaping the role of the designer. Instead of focusing on repetitive and technical tasks, designers have the opportunity to adopt a more strategic and conceptual role, where AI acts as a tool that amplifies their capabilities (Wilson, 2023). This transformation requires not only advanced technical skills, but also a deeper understanding of the values and objectives that each project seeks to convey. According to Martínez and Gómez (2022), this evolution represents a paradigm shift, where the designer goes from being a technical executor to an integral creator who combines creativity, technology and strategy.

AI is forcing designers to reconsider what it means to be creative in a world where machines generate visual ideas. While AI can offer thousands of choices based on predefined parameters, human designers are the only ones capable of interpreting these choices in emotional, cultural and social contexts. According to Sandoval and Jones (2023), "human creativity remains irreplaceable, as machines lack intuition and cultural sensitivity" (p. 34). This rediscovery of human creativity raises new forms of collaboration between humans and AI.

Adapting to an Evolving Environment

Graphic design in the age of AI requires significant change in education and professional development. According to Smith and O'Connor (2023), educational programmes must adapt to include concepts such as machine learning, generative design and data analytics. Universities and online learning platforms are already integrating these topics into their curricula, ensuring that designers are prepared for the challenges of a hybrid marketplace.

In addition, designers must adopt a continuous learning approach to keep up to date with emerging technologies. According to Taylor and Green (2022), "designers who combine technical and creative skills are best positioned to lead in the age of AI" (p. 67).

Artificial intelligence (AI) has radically changed the way graphic design is done, offering tools that optimise creative processes and transform work dynamics. However, this technological advance also brings with it significant challenges and questions about the future of design as a discipline. This chapter focuses on exploring how designers can adapt to this dynamic environment, the perspectives that AI poses for graphic design and a final reflection on its role as an ally in the creative process. The key is to understand that, far from being a threat, AI has the potential to amplify human creativity and open up new opportunities for innovation.

AI is also helping to address sustainability-related challenges in graphic design. For example, optimisation algorithms can reduce the use of resources in printing projects, such as paper and ink, by suggesting settings that minimise waste (Chen, 2022).
Furthermore, in the digital realm, AI can analyse the energy efficiency of designs for electronic devices, ensuring that visual content consumes fewer energy resources without sacrificing aesthetic quality (Brown, 2023). These innovations are aligned with the growing demand for sustainable solutions across the creative industries.

Adaptability is a crucial skill for any graphic designer wishing to thrive in an AI-dominated environment. Today, emerging technologies are redefining the design landscape, and those who manage to integrate these tools into their workflows will not only survive, but excel in an increasingly competitive marketplace (Johnson, 2023). This section explores practical tips and strategies for staying relevant in this new era.

Graphic designers are in a unique position to influence how AI tools are regulated in the creative industry. According to Rodriguez (2022), "designers should not only use these tools, but also advocate for policies that promote transparency and fairness" (p. 22). This includes ensuring that AI-generated images respect copyright and are culturally inclusive.

The emergence of advanced AI tools has led to new specialisations in graphic design, such as generative design and the creation of immersive content for augmented and virtual realities. According to Rodriguez (2022), these emerging areas not only expand creative possibilities, but also create job opportunities in sectors such as entertainment, education and commerce. For example, a designer specialising in augmented reality can create interactive experiences that integrate graphics and elements of the user's physical environment, setting a new standard for visual interaction.

In an ever-changing environment, continuous learning is not an option, but a necessity. The rapid evolution of tools such as DALL-E, MidJourney and Adobe Sensei requires designers not only to learn how to use these technologies, but also to understand how they integrate into their creative processes. According to Fernandez and Lopez (2023), AI and graphic design training programmes are becoming increasingly common in universities and online learning platforms, providing designers with the necessary skills to master these tools.

For example, understanding concepts such as machine learning, natural language processing and image generation can open up new job opportunities for designers. These skills not only increase efficiency, but also allow designers to explore innovative approaches to creating visual content.

Another key aspect of adaptation is learning to collaborate with AI rather than seeing it as a competitor. AI-based tools can automate repetitive tasks, but designers still play an essential role in conceptualising, refining and customising projects. According to Martinez and Gomez (2022), this symbiotic collaboration allows designers to focus on the more strategic and creative parts of their work.

In addition to technical competencies, designers must also develop transversal skills, such as communication and critical thinking. These skills are essential to lead projects involving multiple disciplines and to effectively interpret AI-generated outputs (Taylor & Green, 2022). For example, a designer with strong problem-solving skills can use AI to address complex design challenges, ensuring that solutions are both innovative and functional.

AI is revolutionising lead times in graphic design. According to Perez and Ramirez (2023), AI-assisted design tools can significantly reduce production times by automating tasks such as colour selection, template creation and image editing. For example, Adobe Sensei can analyse an entire project and suggest layout adjustments in seconds, allowing designers to spend more time on conceptualisation and less time on technical execution.

In addition, AI is helping designers work more efficiently in globally distributed teams. Tools such as Figma, which integrate AI capabilities, facilitate real-time collaboration, even between designers in different time zones (Lopez and Garcia, 2024).

The future of AI in graphic design promises exciting developments, but also raises questions about how this technology will influence human creativity. This section looks at current trends and long-term projections to better understand the opportunities and challenges ahead.

Generative design is one of the most promising areas in the use of AI for graphic design. Through algorithms that can analyse parameters set by designers, AI can generate hundreds of design options in a matter of seconds. According to Chen (2022), this not only speeds up the creation process, but also allows designers to explore solutions that would otherwise be impossible to consider.

Another key trend is mass personalisation. AI tools are enabling brands to deliver highly personalised content to their audiences. According to Wilson (2023), this is especially relevant in a world where consumers value unique experiences tailored to their individual preferences. For graphic designers, this means a shift towards adaptive content creation, where a single design can be transformed into multiple variations based on user data.

AI is also playing a crucial role in integrating graphic design with emerging technologies such as augmented reality (AR) and virtual reality (VR). These technologies offer new ways to interact with visual content, enabling immersive experiences that combine graphic design, technology and storytelling. According to Rodriguez (2022), designers who master these tools will have a significant advantage in sectors such as entertainment, education and commerce.

Perspectives of AI in Graphic Design

Constantly updating skills remains a key requirement for designers in the age of AI. Educational programmes are evolving to include concepts such as machine learning, data analytics and generative design in their curricula. According to Martinez and Gomez (2022), 75% of designers surveyed considered AI training essential to stay competitive in the job market.

As AI becomes an integral part of graphic design, ethical and legal concerns also arise. The industry must address issues such as transparency in the use of AI, copyright protection and the prevention of misuse of the technologies. According to Brown (2023), international regulations will be key to ensuring that AI is used responsibly, balancing innovation with ethics.

Rather than replacing human creativity, AI has the potential to amplify it, becoming an indispensable ally in graphic design. This section reflects on how designers can make the most of this technology while maintaining their creative essence.

While AI can automate many tasks and generate visual content, it lacks the ability to understand the cultural, emotional and social context of projects. According to Taylor and Green (2022), this is why human designers will remain essential in the industry. Intuition, empathy and critical judgement are skills that no machine can replicate.

AI is also democratising access to graphic design, allowing non-technical people to create high-quality visual content. This not only broadens the scope of design, but also encourages a greater diversity of voices and perspectives in the industry. However, professional designers must find ways to stand out in a market where access to advanced tools is more common (Chen, 2022).

In the age of AI, technical skills are no longer enough. According to **Fernández and López (2023)**, designers must develop a combination of technological, creative and strategic skills to stand out. Among the skills most in demand are:

- **Data modelling**: Understanding how data feeds AI algorithms can help designers optimise their results.
- **Interdisciplinary collaboration**: Working with software engineers and data scientists will become increasingly common.
- **Ethics and sustainability**: Designers must be prepared to assess the ethical impact of their projects and ensure responsible practices.

On the other hand, transparency will be a key pillar to ensure trust in AI-generated design. According to Taylor and Green (2022), designers have a responsibility to inform clients and the public when using AI tools, and to explain how they contribute to the creative process.

One of the areas of greatest impact of AI in graphic design is mass customisation. Today, brands need to adapt quickly to the individual preferences of their audiences, and AI enables them to do so through data analysis and tailored content creation. According to Wilson (2023), designers who harness these capabilities will be able to lead projects that combine creativity and relevance in real time. This trend is also shaping the way we design for digital platforms, where personalised visual experiences are increasingly in demand.

In addition to personalisation, AI is driving the development of emerging technologies such as augmented reality (AR) and virtual reality (VR). These tools are expanding graphic design beyond traditional boundaries, creating immersive experiences that integrate the visual with the sensory. According to Rodriguez (2022), designers who adopt these technologies will be better positioned to work in sectors such as entertainment, education and commerce, where technological innovation is driving growth.

AI as a Creative Ally

Ultimately, success in the age of AI will depend on the ability of designers to adapt and collaborate with this technology. According to the International Design Association (IDA, 2023), the future of graphic design will be a balance between human intuition and technical capabilities, where both elements complement each other to push the boundaries of creativity.

The challenge is not to compete with AI, but to harness its potential to create more original, inclusive and ethical designs. Designers who master this symbiotic relationship will play a central role in shaping the future of the industry.

The future of graphic design in the age of AI looks like a fascinating intersection of human creativity and advanced technology. Beyond automation and efficiency, AI is redefining the very purpose of graphic design: not just as a tool to solve visual problems, but as a means to connect people in more meaningful and personalised ways. According to López and Hernández (2024), the key to success lies in how designers embrace this technology not to replace their creative vision, but to expand its possibilities and explore uncharted territory.

In this context, AI should not be seen as a competitor, but as a catalyst for innovation. Symbiotic collaboration between humans and machines can open doors to new forms of visual expression, integrating values such as sustainability, diversity and inclusion. This means that the future of graphic design will not only be defined by the capabilities of technology, but also by the intention and values that designers choose to prioritise. AI not only amplifies creative capacity, but also presents the opportunity to rethink the boundaries of what graphic design can achieve in an interconnected and ever-changing world.

Conclusions

The New Art of Creating: How AI Redraws Graphic Design highlights how artificial intelligence transcends its technical role to become a catalyst that amplifies human creativity. The book analyses how AI automates tasks, optimises processes and extends the capabilities of designers, while raising reflections on ethics, authorship and sustainability in graphic design. Beyond its technological advances, it emphasises that the value of design lies in how professionals integrate these tools to enhance their artistic vision and lead responsibly.

The text recommends adopting a continuous learning mindset, balancing the use of AI with ethical practices and developing strategic skills to lead interdisciplinary projects. In short, the book argues that AI does not replace human creativity, but rather amplifies it, positioning itself as a key ally in building a future where innovation and design merge into meaningful and transformative solutions.

Bibliographical references.

Abadía, J. A. (1986). *Historia de la creatividad.* Ediciones Visual.

Academy Partners (2023). Adobe Sensei and AI in design. Academy Partners.

Academy Partners (2023). Key competencies in the age of AI. Academy Partners.

Adobe (2022). Adobe Sensei: Reinventing creative experiences with AI.

Brown, J. (2023). Ethics and AI in graphic design: Challenges and opportunities. *Journal of Digital Creativity, 10*(2), 45-61.

Brown, L. (2023). Intellectual property and AI in design: Challenges and opportunities. *Creative Journal, 45*(2), 112-118.

Cely, C., & Buake, J. (2023). Creative collaboration between humans and AI: A graphic design perspective. *Design Thinking Journal, 15*(2), 125-137.

Chaparro González, L. (2024). *Development of a virtual reality demo with tools from IA*. Final degree project, Polytechnic University of Catalonia.

Chen, H. (2022). Cultural diversity in AI design. *Journal of Design Ethics, 10*(3), 45-56.

Chen, L. (2022). Generative design: Pushing the boundaries of creativity. *Creative Tech Journal, 18*(2), 99-110.

Deloitte (2020). *The impact of artificial intelligence in the workplace*. Deloitte Insights.

Ethical Design Initiative (EDI) (2023). *Ethical guidance for AI-driven design.* EDI Publications.

Fernández, M., & López, P. (2023). The AI revolution in graphic design. *Innovation Press.*

Gómez, A., & Pérez, L. (2021). Legal frameworks for AI-generated content. *Law and Innovation Journal, 8*(4), 56-72.

Gómez, R., & Rodríguez, M. (2023). Effects of AI on visual creation. *Journal of Technology in Design, 5*(4), 125-132.

Goodfellow, I., Pouget-Abadie, J., Mirza, M., Xu, B., Warde-Farley, D., Ozair, S., ... & Bengio, Y. (2014). Generative adversarial nets. *Advances in Neural Information Processing Systems, 27*, 2672- 2680.

González, M. (2020). Deep learning for image manipulation: Opportunities and challenges. *Journal of Visual Computing, 23*(4), 145-158.

Hernández, J., & Torres, M. (2024). Artificial intelligence and its integration in creative workflows. *Design Futures Quarterly.*

AI Tools (2023). *A practical guide for designers in the age of artificial intelligence.* Creative Tools Publishing.

Johnson, K. (2023). Future skills for designers. *Design Studies Quarterly, 15*(4), 78-89.

Johnson, R. (2023). Advanced prototyping: How AI is changing industrial design. *Industrial Design Today, 9*(1), 45-57.

Kaur, M. (2023). Collective creativity: Humans and machines working together. *AI & Design Review, 12*(3), 88-102.

Lazo Altamirano, J. E., Condori Quispe, M. Y., & Abarca Rojas, R. J. (2024). Impact of artificial intelligence in graphic design. *Journal of Scientific Research KUTIMUY, 12*(2), 100- 109.

Lee, J. (2022). The challenges of data usage in AI training. *Technology and Ethics, 7*(1), 33-45.

Lee, J., & Park, H. (2023). Technology training for designers of the future. *Global Design Survey*.

López, J., & García, M. (2024). The cultural impact of AI in design. *Art and Technology Review, 19*(2), 34-50.

Martínez, R., & Dennis, T., Cáceres, A., Gutiérrez, L., & Quiroz, J. (2023). The impact of AI tools in modern design workflows. *Creative Innovation Journal, 14*(1), 518-532.

McKinsey & Company (2023). *Redefining creative teams in the age of AI*. McKinsey Digital Report.

Miller, P., Evans, D., & Carter, L. (2023). AI-driven immersive environments: Virtual and augmented reality in design. *Immersive Design Quarterly, 7*(2), 30-47.

Pérez, S., & Ramírez, J. (2023). Algorithmic creativity: Redefining art and design in the AI age. *Art & Technology Review, 11*(4), 34-48.

Rodríguez, A. (2022). Bias and inclusivity in AI-driven design. *Ethics in Media Journal, 8*(2), 78-92.

Rosenberg, D. (2021). Machine learning in graphic design: Transforming creativity. *Design Technologies Journal, 4*(1), 23-39.

Sandoval, R., & Jones, T. (2023). Ethics in algorithmic design: Prospects for the future. *Ethical Design Studies, 11*(1), 22-39.

Santos Tapia, F. D. (2024). Automated Graphic Design: A critical analysis behind the artificial intelligence. *Eídos, 24*, 81-93.

Schwartz, P. (2021). The history of artificial intelligence in design. *New Horizons in Design,*
10(1), 10-24.

Smith, A., & O'Connor, B. (2023). Towards an international ethical charter for AI design.
Global Ethics Council.

Taylor, J., & Green, K. (2022). Transparency and accountability in AI design. *Ethical Perspectives, 11*(3), 22-35.

Taylor, M., & Green, E. (2022). Ethical considerations in AI-driven design. *Journal of Design*
Ethics, 15(2), 23 37.

Veritas, A. (2023). Creativity in the AI age: Challenges and opportunities. *Design Futures*
Publishing.

Wilson, P. (2023). *AI as a creative collaborator*. New York: Design Horizons Press.

Wilson, P. (2023). *Inclusivity through AI in design*. Inclusive Technology Review, 18(5), 66-
78

Printed by Books on Demand GmbH, Norderstedt / Germany